ALSO BY GARY TAUBES

Why We Get Fat: And What to Do About It

Good Calories, Bad Calories: Challenging the Conventional Wisdom on Diet, Weight Control, and Disease

Bad Science: The Short Life and Weird Times of Cold Fusion

Nobel Dreams: Power, Deceit, and the Ultimate Experiment